CARE AND REPAIR OF ANTIQUE METALWARE

Chris Browning

D1146149

TIGER BOOKS INTERNATIONAL
LONDON

To my parents who made me understand
that craftsmanship is made of hand plus brain
but above all heart

CLB 4488
This edition published 1995 by Tiger Books International PLC, London
© 1995 CLB Publishing, Godalming, Surrey
All rights reserved
Printed and bound in Singapore
ISBN 1-85501-637-0

Note: Throughout this book, American terms are signalled in
parentheses after their British equivalents the first time in each
section they occur.

Editors: Maureen Maddren, Eric Smith and Anne Yelland
Editorial consultant: Nick Russel
Editorial assistant: Catherine Tilley
Art director: Elaine Partington
Art editor: David Allen
Designers: Nick Maddren and Su Martin
Illustrators: Graham Bingham and John Woodcock
Photography: Jon Bouchier
Studio: Del & Co
Picture research: Liz Eddison

Contents

Introduction

Ancient skills

The skill of the metalworker is almost as ancient as mankind. No other material has had such an effect upon life the world over. It still retains, for the uninitiated, something of the black art once associated with the alchemist or magician. The working of metal, especially for weapons or body adornment, is still seen in certain cultures as an act to be passed on to a selected few. This has nothing to do with the muscle power required to work the material but to the power the maker holds over the community.

Metal is to most people, even today, a relatively unexplored material. In the average household, while there is likely to be a basic woodworking kit – even if it is just a hammer and a saw – its metalworking equivalent is usually absent.

Although there are many metal objects within the home they tend to be replaced rather than repaired once they become damaged. This is not so with antiques or decorative objects, which is why the particular pieces illustrated in this book, rather than articles in everyday use, were chosen. There is no reason why more metalworking should not be undertaken by the amateur providing it is approached in the correct manner.

The average person treats metal as a totally alien material. It can, however, do most things if treated in the correct fashion. The main thing to remember is that any action you take against it tends to be fairly permanent. The approach to metalworking, therefore, has to be thorough and methodical. That is not to say there is only one way to achieve your objective. This book is meant to guide and encourage you to explore your own potential on the material and your own skill.

The examples shown are not intended to be step-by-step techniques guides: a particular technique on one example may be ten times more difficult on the next piece you try. Rather, they are there to show what can be achieved with relatively simple equipment and care. Hopefully, after reading the book you will be able to restore your own pieces with an added degree of confidence.

Left: *Many repairs to domestic metalware require only basic tools: here, a vice secures the stake supporting the kettle during planishing* (top) *; a gas torch heats the lamp cap until the solder is liquid and the retaining tabs can be removed.*

What can be achieved?

Ground high tensile steel wire is used to ream out a blocked hole in this policeman's lamp.

When deciding whether to restore an object you should be clear why and how you are going to do it. The restoration of an article will be affected by the use the owner wishes to put it to.

Do you repair it to the best of your ability, regardless of damaging its resale value? Do you use modern materials because they are designed to do just the job you need? When replacing parts should you make sure they blend in or leave the repair obvious?

It is not skill alone that needs to be developed but an appreciation of the effect your work will have on the final piece. Traditional tools and working methods all help to create a climate where an understanding of the reasons behind the existence of a piece can flourish.

All the items you work on were made for a good reason. It is worthwhile doing a degree of investigation into the history of a piece before you start work. Try to understand how a piece was made and in what order. Frequently a repair will require the removal or dismantling of a large part of the object. An insight into its original manufacture can save hours of frustration, struggle and possible damage. Learn to recognize materials and the techniques used on them. Remember most items are made from a particular metal because that metal is good at the job: either the use to which the item will be put or its performance in a particular process.

Ask yourself why the item was made in this material and why it was made in this shape.

There are, in theory, no limits to a piece which can be restored. Often an amateur craftsman can afford the time to undertake a repair which would not be feasible to a professional due to the economics involved. Do not be put off by lack of experience but build on your skill, sensitivity and care. As in early schooldays, the novice craftsman must learn the important Rs:

Recognize the type of metal, the techniques required, the period of the piece.

Record and store the information obtained in a manner in which it can easily be referred to.

React to the information you have gathered and choose your course of action.

Remove any damaged elements from the main body of the item, as well as any material not helpful to a successful repair.

Remake/repair after sufficient investigation, the choice is yours.

Re-assemble the item exactly as it was – especially if it is mechanical. Refer to your records.

Re-finish trying to use materials which are contemporary to the piece. If possible, retain original patinations.

Rewards could be financial or just a 'Well done'.

Many pieces of domestic copperware can be restored to their former glory – all that are needed are care, patience and a sympathetic approach.

Materials

Emery cloth is a natural abrasive, blue/black in colour, mounted on cloth and used for cleaning and polishing. Several grades are available. Do not use on silver, because it can burrow into the surface.

Steel wool can be used with a little oil to clean a corroded surface or to give a matt finish to an article.

Crocus papers are used for fine polishing by hand, several grades being available. They provide an alternative to Water of Ayr (Scotch) stone for fine scratch removal and will produce a high polish.

Methylated spirits (wood alcohol) is useful for cleaning surfaces before soldering. It is also mixed with fire-stain inhibiter (see Argo-Tec below).

Paraffin (kerosene) is a general-purpose cleanser. Used with steel wool, it will clean badly corroded surfaces.

Beeswax is used in block form as a lubricant for piercing saw blades. Mixed with turpentine it will protect freshly cleaned ferrous surfaces against rust.

Thrumming thread is used in hand polishing and is charged with tripoli or rouge. Tripoli is used in hand or machine buffing for surface scratch removal before using rouge. Rouge is used in powder or bar form for the final polish.

Pumice powder is mixed with water or oil to make a controllable slurry which can be used to clean metals before polishing or after heat treatment.

Water of Ayr (Scotch) stone is a very fine, slate-like material used in the removal of small scratches. It should always be used with plenty of water and with care, as it can very easily form a hollow on the metal.

Fluxes used for soft soldering include zinc chloride which is an active (corrosive) flux and any residue must be cleaned off after use. It must not be used on electrical work. Ammonium chloride in petroleum jelly is also classed as an active flux. Again, any residue must be removed. Tallow and resin are 'safe' fluxes for soft solders, but will leave a greasy deposit. Active fluxes are usually more effective as their slight acid action helps to clean the metal. It is important that any residue of active flux is removed after soldering as it is so corrosive.

For hard soldering, borax can be used as a flux for most silver soldering but there are proprietary fluxes available which are sometimes more suitable as they are made to be used with a specific solder grade; Argo-Tec is a powder used for the prevention of fire stain when hard soldering silver. It is mixed with methylated spirits to form a creamy paste.

1 Selection of emery paper; 2 rouge; 3 thrumming thread; 4 borax cone; 5 beeswax; 6 Water of Ayr stone; 7 steel wool; 8 methylated spirits; 9 paraffin; 10 solder.

Tools

Special tools for metalwork tend to be very expensive, but if they are looked after they will last for decades and will prove to have been an excellent investment. It is said that a bad workman always blames his tools but nothing makes a job easier than having the correct tool that is in good condition. Many repairs can, however, be undertaken with a very basic toolkit and it is possible to modify some of the more common items to perform a specialist function.

Building up a comprehensive toolkit can literally take a lifetime but you do not have to spend a fortune on equipment to achieve good results. A toolkit comprising the following basic items will allow you to carry out many of the more straightforward repairs, and you can add specialist tools as and when the need arises.

Basic tools

A small vice (vise) is ideal for holding a piece securely while it is being repaired. Vices come in all price ranges and can be either dismountable or permanently fixed to the bench. The type which has a swivel head is a great help in getting at awkward angles. Make sure that the type you choose is fitted with soft jaws which will not damage the work.

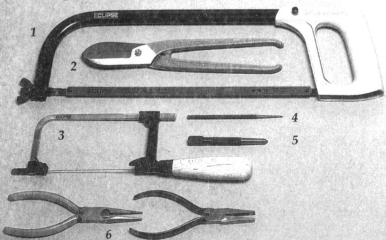

1Hacksaw;2tin-snips;3piercingsaw;4scriber;5centre punch;6pliers;

A gas torch is invaluable as a source of heat for soldering, loosening jammed parts, hardening and tempering (see p. 22). If more than the occasional repair is to be attempted, a torch with a separate gas tank is an excellent investment. Always get a torch or heat supply with a greater capacity than you initially think you need. Reserve capacity, when soldering for example, makes for speedy and efficient working. Remember you can always reduce the amount of heat but you cannot increase it if the equipment is unable to supply the demand. Larger torches have the added advantage of being able to provide varying sizes of jet so that flame shape and size can be changed as necessary.

Miscellaneous tools such as the following are recommended: a hacksaw (full-size or junior), pliers of various sizes, tin-snips, screwdrivers, a wooden mallet, a ball peen hammer, marking-out equipment (scriber, centre punch, square and dividers), an assortment of files, an electric soldering iron and a jeweller's frame saw, which is probably the most important of metal-cutting tools. Be sure it is of good quality; the most convenient type has an adjustable frame.

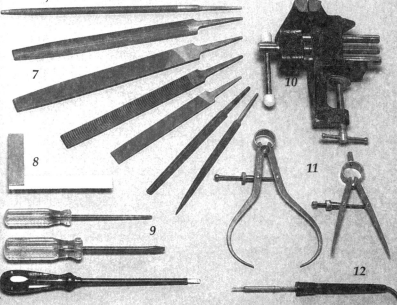

7files;8square;9screwdrivers;10vice;11dividers;12soldering iron.

Specialist tools

A planishing hammer is a good investment, although a standard carpentry hammer can be used. Planishing hammers come in various weights but a general-purpose hammer would be between 140g (5oz.) and 280g (10oz.). They must be kept highly polished and protected when not in use.

Clamps can be made of metal or wood and usually have a simple closing system of a metal slide or wedge. They are used for holding small pieces of work while you are filing or polishing and are used in conjunction with the bench pin.

The bench pin is a hardwood V-shaped tool used to hold irregular items when piercing with a saw or filing to shape.

The pin vice is ideal for working on wires or pieces which have slender shanks. They come in various sizes and can also be used to hold very small drills.

The scraper is ideal for taking burrs off metal edges before soldering or taking material off the surface of the work piece. Most useful to the metalworker is the three square hollow ground type. It is important to keep it well honed.

The burnisher is used to smooth and consolidate the surface of the metal into a high polish. It is made from hardened and tempered high-carbon steel.

Mallets: four main types are used – round boxwood, rawhide, bossing and soft-faced.

A scoring tool is made from high-carbon steel which is hardened and tempered (see p. 22). It is used to cut a groove of specified angle in sheet material. The sheet can then be folded up and soldered in place.

Needle files are used for fine intricate work with the minimum of material removal. Each file has a handle and there are a variety of sizes. They should be purchased individually and not in a pack. The most useful shapes are the hand, warding, three square, half round and round; there are eight other shapes. The cuts range from 00 which is the coarsest to 8, the finest.

Swiss files are slightly different from files normally purchased in the UK or the United States. They are made from chrome steel for durability and hardness. They have a range of shapes which is a little different from their British and American counterparts and, in addition, they have slightly more taper. They are designed for precision work and should be treated with great care.

Rifler files come in many shapes. They are used to file in areas which would otherwise be inaccessible. Both ends of any one tool have the same shape.

File care is vital. Files used on non-ferrous metals should not be used on steel. If you are working very soft alloys (e.g. aluminium-based or lead-based), keep separate files for this. Particles of lead transferred by file to a piece of silver can cause untold damage. Good files should be kept separated and not allowed to clank together. Do not use oil on files as this impedes the clearance of material. Files should be cleaned using a piece of copper which is worked obliquely across the surface to remove the small, hard pieces of compacted waste material known as 'pinning'.

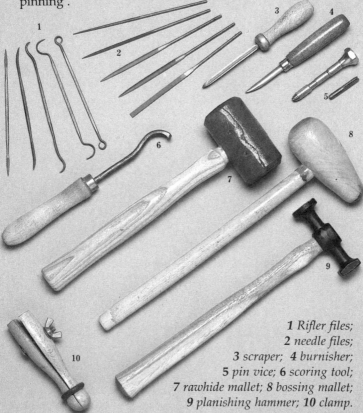

1 Rifler files;
2 needle files;
3 scraper; 4 burnisher;
5 pin vice; 6 scoring tool;
7 rawhide mallet; 8 bossing mallet;
9 planishing hammer; 10 clamp.

A place to work

Anybody interested in craft work would love to have a custom-made workshop. Unfortunately this is available only to a few lucky individuals. For those of us with more limited means and limited space, all is not lost. You do not need wonderful facilities to produce first-class work but you do need to be able to carry out work with the minimum of upheaval. Any activity, whether it is a pastime or a profession, soon becomes a chore if a lot of preparation is needed before you can even start.

Your workspace need not be permanent but it should be accessible. Always try to plan your area no matter how small it may have to be. You may find that you need separate areas for clean work and dirty work. Divide your activities into main categories such as cleaning, fabrication and finishing, then consider the implications of work to be carried out under these headings. For example: Is water to be used? Is oil to be used? Will there be any noise/vibration, dirt, smell, etc? Is the vice (vise) big enough for filing? Is the area firm enough for hammering? Will this cause an excessive amount of noise? Hand polishing usually entails little or no problem. Is there a power supply for machine polishing to take place? Is the area going to be contaminated with dust?

By considering these factors, you can soon work out what sort of area you will need. If a proper workbench is not available, a table-top model can be used. This just slips on to the table rim and can be put away when not in use. Refractory surfaces can be screwed to the top when soldering is to be carried out and additional items like racks for tin-can storage can be added as you wish.

No matter how restricted your area do not skimp on light, warmth and ventilation, which are fundamental to achieving lasting success in your work. And be sure to pick a comfortable chair which will support you while you work and not put any strain on your back; this will make the job a lot easier and more enjoyable. When sawing, however, use a low stool so that the saw is at shoulder height.

Right: *An easily made work surface which can be used on an existing table top. It can be removed for storage when not required.*

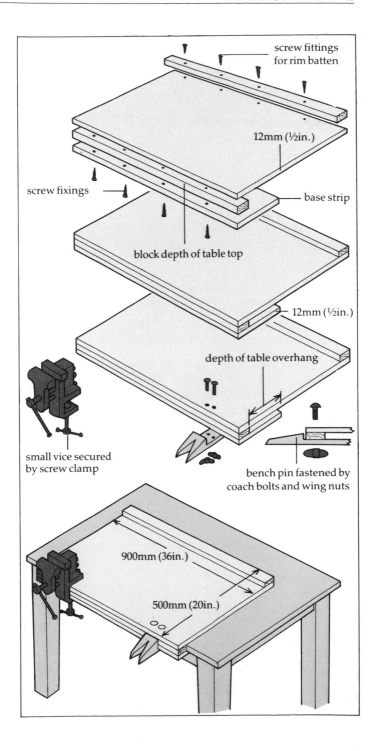

screw fittings
for rim batten

12mm (½in.)

screw fixings

base strip

block depth of table top

12mm (½in.)

depth of table overhang

small vice secured
by screw clamp

bench pin fastened by
coach bolts and wing nuts

900mm (36in.)

500mm (20in.)

Metals in common use

When you are dealing with a repair on any piece of metalwork, a successful conclusion to the task can depend on your ability to identify the materials involved. In most cases this will be fairly obvious but it is always worth that second look. If you have any difficulty, it is usually the original method of manufacture which can provide the all-important clue.

Different metals are used for different reasons, basically because one is known to be better suited to a particular requirement than another. Cost will also have been an important factor, so always try to judge a material from its looks, its purpose and the overall quality of the piece. Mistakes can be avoided by a little extra care. Some of the more common metals, along with their descriptions, are listed below.

Ferrous

Cast iron is used in heavy sections; it takes compression but does not like shock loads. Will take fine detail when cast but is difficult to repair. When ground, non-bursting dull red sparks are produced.

Mild steel is available in all forms from wire to sheet. Will rust if unprotected. Will not hold a cutting edge. Light silvery-grey when clean; will polish to a high finish. Will appear in numerous components. Sparks are bright yellow; a few are star-like when ground.

Tool steel has a high carbon content and a grey, mottled appearance when hardened. Takes a high finish. Used for tools, cutting edges and guides. Produces a stream of bursting, star-like sparks when ground.

Wrought iron is used on good-quality scroll-work. Forges easily and works readily when red-hot. Has a distinctive surface when etched.

Non-ferrous

Aluminium alloys are light in density. Colours are off-white to silver. Difficult to solder. Most will anneal if carefully heated. Very soft in the main; easily cut. They resist corrosion but can be etched with certain acids.

Brass is an alloy of copper and zinc. Many types, but all are yellow in colour. Stiffer than copper after heating but easy to machine. Takes a high polish.

Bronze is an alloy of copper and tin. An ancient metal, harder than copper, it casts well, unlike copper. Good resistance to wear and corrosion. Many types for special uses; the type used for art casting is often found with a brown patina.

Copper is reddish/gold in colour. Comes in all forms: wires, sheet, foil, pressing, spinning and so on. Solders well, softens with heat; can be cut and machined. Malleable and ductile.

Gilding metal is made from brass with a very high copper content of 80-95%. Golden in colour, used for decorative work and imitation jewellery. Used by silver-smiths for prototypes. Has closest work characteristics to silver. Will soften with heat but stays stiffer than copper.

Lead is heavy, soft, bluish-grey, and has low melting point. Cannot be work-hardened, surface oxidizes quickly. Traditionally used as a base in old pewter. Used for garden statuary. Mixed with tin to form soft solder.

Modern pewter contains no lead, unlike old pewter. Dull-silver in colour, can be soldered. Soft, easily cut.

Nickel silver (also known as German silver or green silver) is an alloy of copper, zinc and nickel. Much tougher and springier than gilding metal. Retains stiff-ness when heated; pale greenish-silver in colour. Used for hinge pins, hinges, cutlery.

Pewter is 94% tin, 4-5% antimony and 1-2% copper; known as Britannia metal.

Silver is white to silver in colour, will take all mechanical processes; solders well and is malleable and ductile. Used for jewellery, electrical work, decorative metal-work. Plating on cutlery or areas subject to corrosion.

Tin is soft with a shiny, white surface which stays bright owing to very good anti-corrosion qualities; very low melting point.

Zinc is blue-grey, soft, used for protective layer on steels. Zinc alloys used for many types of castings; these can easily be mistaken for other metals. Often requires mechanical test to identify.

Techniques

Soft soldering
This technique is dealt with on p. 44.

Hard soldering
Hard solders have been in use for many years. They appear in articles ranging in scale from jewellery to engineering. There are many types to choose from, most being based on silver or gold. When soldering precious materials it is important to use only solders which will not debase that article.

To ensure that no debasement occurs, silver solder conforms to sterling standard. There are, however, cheaper alloys available for use when working on base metals such as copper, gilding metal and brass.

Hard solders are alloyed so that the join is not betrayed by a 'colour line'. Therefore solder for use on a gilding metal, for example, would be much more yellow than the equivalent grade for silver. It is crucial that solders for base metals and those for precious metals are not allowed to become intermingled.

Although they all come under the general heading of 'hard solder', there are five basic grades which are ranged according to melting point. They are:

Extra easy	680-700°C (1256-1292°F)
Easy	705-723°C (1301-1333°F)
Medium	720-765°C (1328-1409°F)
Hard	745-778°C (1373-1433°F)
Enamelling	730-800°C (1336-1472°F)

You will notice that almost all the melting ranges overlap. Each manufacturer has his own form for these solders, so they may be supplied as round wire, for example, or as flat strip. Get to know the various cross-sections and the grades they represent.

Hard solders are poor at bridging gaps. Surfaces to be joined must be really close-fitting along their whole length and width. This may mean that edges have to be filed, or possibly that old solders have to be scraped off.

Ensure that the surfaces are perfectly clean because

solder will not flow along dirty joints. Parts can be pickled (immersed in a dilute acid solution) and then washed to remove oxidation, or they can be scoured with pumice.

Once they are fitting snugly together and are free from grease, the parts may need to be held in place. How this is done will depend upon the article being worked on. They can be wired together using iron binding wire, or held by a small weight. Alternatively they can be kept in position by using split pins. If these are heated and quenched, they are ideal for holding fast pieces which are being soldered: as the pins have been heated, the oxide layer created will act as a solder inhibitor and prevent them from becoming permanently attached. In addition they will be 'dead', which means that any subsequent heating – as during soldering – will have little effect on them. Pieces can also be held in place by 'stitching', which is raising a burr up by the use of a graver.

Methods of securing while soldering

1 *Soft iron wire will keep joints closed.*

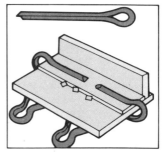

2 *Split pins raise piece clear of hearth floor.*

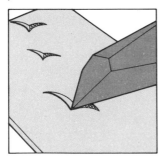

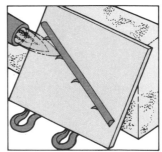

3 *Stitches to hold small pieces in position can be made using a graver.*

Fluxing and soldering

The heat required for soldering would normally cause oxides to form on the surface of the metal. As oxides prevent solder from adhering, it is necessary to protect the surface from air contact. This is done by applying flux (see p. 9) to the areas to be joined.

Flux has several functions:
  i) it will break down any thin surface oxide
  ii) it prevents further contamination
  iii) it will reduce the surface tension of the solder and assist it to flow along the joint.

Flux can be purchased as a powder, a liquid or in cone form. A thin, creamy paste can be made by dissolving the powder in a little water or by rubbing the cone in water in an unglazed dish.

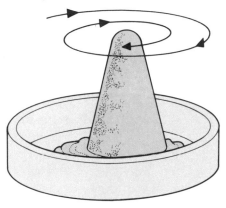

Rubbing the borax cone in a little water in an unglazed dish produces a thin, creamy paste flux.

Paint the flux on the joint. The heavier the pieces to be soldered the thicker the flux required. Too thin a deposit will burn out before the solder melts. Too much will cause the solder to spread over the surface rather than along the route of a joint.

Once the piece has been fluxed and secured, heating can take place. Remember to raise the objects off the hearth slightly to enable the flame to heat all parts evenly.

As the temperature rises, the flux will bubble. Watch carefully and make sure that the bubbling has not displaced any parts. With additional heating the bubbling will subside and the flux will form a jam-like deposit over the joint. This prevents air from reaching the joint and forming an oxide. Heating continues and eventually the flux will melt and run across the surface and into the joint. It is at this stage that the solder is

Soldering

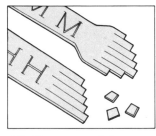

1 Scratch the grade on the solder sticks.

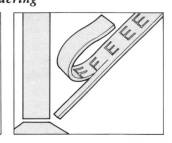

2 Solder sticks are reduced in width when 'fed' in.

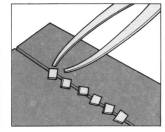

3 Paillons are made by piercing small sections.

4 Place paillons on the joint with tweezers.

introduced. This sudden change of state of the flux is a perfect indication of the required temperature and should be looked for. The solder can be fed into the joint as a stick or pre-placed along the joint as paillons panels, which are small pieces of solder cut from the strip. If you use these small chips, you will find it easier if the job is first heated just enough to let the flux bubble up and die down. Coat the chips with wet flux and place them in position with tweezers. The flux on them immediately dries and holds the chips in position.

Only the minimal amount of solder should be used. If it does not flow, it is because you have failed to give it the correct conditions for movement. Make sure that both parts of the job are at the same temperature or the solder will be attracted to the hotter one. Try to get the solder to melt by using the heat transferred through the job rather than by playing the flame on the joint. Excessive heating of the solder will make the zinc in the alloy vaporize and cause pin-holes in the joint.

Finally let the job cool before pickling or quenching. Premature quenching only puts stress on the joint.

Hardening and tempering

At some time it will be necessary for you to make a small tool either for a special job for which you cannot buy the tool you need, or because the one that you need is no longer commercially available.

Items such as punches, chisels, scoring tools and scrapers are best made to the individual requirements of the craftsman and the ways in which he works. They are nearly always made from high-carbon steel (tool steel) or silver steel. In order to function properly, good tools must be hardened and tempered to suit the use to which they will be put.

By heating tool steel until red and then quenching quickly it will become very hard. In this condition, however, it is too brittle to be of any practical use. A secondary heat treatment is required to reduce the hardness a little but increase the toughness. This is known as 'tempering'.

When making a tool, the desired shape is achieved by sawing and filing. The surface finish needs to be as smooth as possible. This results from using progressively finer grades of emery cloth.

High-carbon steel will usually be hardened in clean water but the addition of salt to the water increases the degree of hardness, as the heat is conducted away more quickly. Care must be taken to prevent cracking, which can be caused by overheating and can occur in both clean and salt water. Cracking will follow a weakness on the surface, hence the need to polish out any scratches before hardening. Hardening should be carried out on a rising temperature, as this will produce a finer grain structure in the metal.

When quenching be sure to plunge the tool vertically, otherwise it is liable to bend. Stirring the water before plunging the tool will take away any steam jacket formed on entry.

Hardening will produce a characteristic mottled surface composed of oxide, all of which must be polished off with fine-cut, dry emery cloth before tempering. Do not use oil as this might leave a light film which will affect the next stage. Do not touch the hardened end as oil from the skin will also detract from the final temper. Hold the tool in a small flame about 25mm (1in.) behind the hard end. As it heats up, oxide will form on the steel. The rise in temperature will be accompanied by a change in colour, according to how hot it is.

These changes are:

Pale yellow	220°C (428°F)	High temper
Pale straw	230°C (446°F)	
Middle straw	240°C (464°F)	
Dark straw	260°C (500°F)	Medium temper
Purple	270°C (518°F)	
Blue	290°C (554°F)	Low temper
Pale blue	320°C (608°F)	

The heat will travel along to the tip, forming bands of moving colour. Once past blue, oxides continue to form but by that time the steel is too soft to be of any use.

When the tip of the tool is at the desired colour, it is plunged into oil. The degree of hardness is then captured at that point. To decide the colour required it is necessary to refer to a tempering chart.

Pale yellow	engraving tools, files, light turning tools
Pale straw	razors, scrapers, twist drills
Middle straw	cold chisels, hammer faces, penknives, taps, dies, punches, shears, reamers
Dark straw	wood cutting tools, chisels
Purple	wood boring tools, axes, needles, springs
Blue	springs, screwdrivers

The surface colour can either be polished off or left to enhance the appearance. If in the process of tempering the colour required is passed by, then the whole process of hardening and tempering must be gone through again from the beginning.

Old files are a very good source of good quality steel, but they must be softened thoroughly before they can be used. Heat slowly until they are red, then allow them to cool. By letting the temperature drop as slowly as possible all the internal stress in the steel caused by hardening will be released and it will change back to a softened state.

Polishing and finishing

Polishing and finishing are the last stages in the repair and restoration of any article. The surface may be left matt or worked to a high polish.

The principle of polishing is to start by dealing with the worst marks using the coarsest abrasive; follow successively with finer abrasives until the finest is reached. Excess solder and deep scratches will probably need a file or emery cloth to remove them. Fine scratches are then removed with Water of Ayr (Scotch) stone or paper-backed abrasives.

Once any residue has been washed away, the surface should be perfectly matt. Jeweller's rouge can be used for the final polish. This can be in cake form (bound with hard grease) or in powder form which is mixed with water to form a paste. The rouge can be applied with a stiff brush or chamois leather. Any pierced work can be polished by thrumming thread. A few short lengths of thread are tied to a hook at the bench front and pulled taut. Wipe the rouge along them, pass them through the pierced work and rub the piece back and forth. Threads will quickly cut a groove so make sure to work over the whole surface evenly. Finally, wash the piece in warm water with a little added detergent.

Quicker results can be achieved by using a polishing motor and buffing wheels. These vary in diameter, thickness and composition. When used in conjunction with the correct polishing compound, each wheel will give a different effect, the general rule being that the harder the wheel the coarser will be the finish obtained. It is important that wheels are kept for use with one grade of polish and do not become contaminated by a different grade. The type of abrasive used can be marked on the mop side. After polishing with one type, make sure that the article is cleaned before moving on to the next wheel and abrasive grade.

Motors for polishing have one fixed speed, around 3,500 r.p.m. Polishing speed will depend upon the diameter of the buffing wheel, which is expressed in surface feet per minute (s.f.p.m.).

The recommended s.f.p.m. will depend upon the metal being finished, the kind of wheel, the type of abrasive and the finish required. What is important is that there are no short cuts to a good finish. Good preparation of each component will repay the time spent many times over. When buffing, keep the work on the move constantly otherwise surface damage can occur. This is especially true in the case of hard wheels, which are normally used for flat surfaces and to prevent blurring square edges, but they must be used with care.

Soft wheels are used where the surface is irregular and it is not intended to have fast metal removal.

Care of buffing wheels

From time to time it is necessary to clean the mops as they become clogged with unused abrasive and re-moved metal. This can be done by holding against the wheels a length of wood with protruding nails. Alternatively use a long-handled, stiff-wired brush. Do not try to wash the wheels as this can break them up and reduce their effectiveness.

Wheel types

Below is a short list of wheel types and abrasives. Always use the utmost care when polishing. Make sure that any loose clothing or long hair is kept confined. Finger rings, if worn, should be removed if possible or taped over to prevent snagging.

Soft wool (for soft metals) gives a fine finish; use with rouge.

Flannel and cotton types have a nap on the fabric. When used with rouge, they will not scratch.

Muslin wheels vary between light, loose discs of fabric to closely woven, heavy starched canvas. They are either unsewn or can be stitched in a variety of geometric patterns. A number of compounds can be used with them; tripoli is the most common.

Canvas is used for the hardest cloth wheels for fast metal removal. The canvas can be cemented together and produces a harder wheel than stitching. Use with tripoli.

Pressed felt wheels come in grades ranging from soft to rock-hard, some wheels being pre-shaped for access to awkward areas.

Bristle brushes are used with greaseless brushing compounds to create a matt surface.

The three most common types of abrasives are tripoli (the coarsest), crocus (intermediate) and rouge (for the final polish). There are many types of each of these.

Project 1: Scales

First impressions

On first inspection these scales appear to be incomplete. They do not have a stand of any kind and comprise solely the balance beam with pointer and the pans. However, there is no provision made for any additional pieces and it would appear that they are to be hand-held, probably from the finger inserted through the top ring. Alternatively, they could have been suspended from a hook although the amount of wear on the ring is minimal.

The quality also is better than a cursory glance would indicate. The pans, although they appear to be rather thin and cheap, in fact are on the heavy side for their size. This is due to the manner in which they have been made. They have been made out of a sheet of copper (in itself unusual, as most scale pans are of brass). The material apart, the manner of their forming is out of the ordinary. They have been raised from the sheet. This process is now only used on very good-quality hollow-ware, and is an aid to putting a date on an otherwise common everyday object. If they had been made within the last century there would be a good chance that the pans would have been made by some semi-automated process: either by stamping or possibly by spinning. Also the bowls are quite deep in relation to their diameter. This would indicate their use as a commodity balance rather than a weighing scale.

The beam is also an indication as to the age and quality of the item. It is made from wrought iron and has been forged in one piece. When cleaned, it shows the water pattern markings common to this material.

Order of work

Although generally in a reasonable condition, one of the scale pans has quite a severe dent in it which will have to be removed. Before this can be attempted, however, the chains must be removed to facilitate access to the pans, and the pans must be cleaned. These scales have been lacquered at some time in their life and this must be removed before hammering begins.

Once the pans have been restored, the chains can be cleaned and polished, and the balance beam cleaned of rust.

Left: *The damaged scales, with the pans dented and creased; and restored and reassembled* (top).

Hollowing and sinking

There are several ways to make a hollow vessel out of non-ferrous material by hand. The technique used depends upon the depth and diameter of the object and also the quality.

The first method 'hollowing' (or blocking) is used for making bowl shapes. It can be done with a wooden mallet or metal blocking hammer. The metal is shaped by hammering it into a depression in either a wooden block or sand bag. It is gradually worked into the depression by regular hammer blows starting from the outside edge and spiralling into the centre, each successive row being the width of one hammer blow.

If the vessel is to have a flat rim the hammering will commence at a distance in from the edge determined by the rim width. This is referred to as sinking.

The main difficulty with this technique is in trying to keep the base of the article flat.

Hollowing and sinking

1 *Start at the rim to hollow a bowl.*

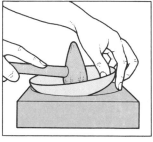

2 *Support the back rim with your fingers.*

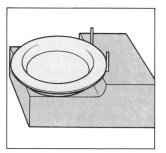

3 *Sinking is similar to hollowing.*

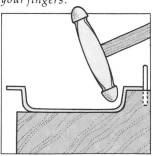

4 *Rotate the work as hammering progresses.*

Raising

Most items are too narrow and deep to be made just by blocking alone and these will have to be raised. Both in sinking and blocking the object is hammered from the inside; in raising it takes place on the outside of the form. The process requires the use of a metal bar or stake against which the metal disc is worked. Starting from the base the metal is hammered in concentric circles pushing it on to the stake. Raising makes a characteristic rectangular flat mark on the metal which is later removed by planishing. With this process the metal would need to be softened (or 'annealed') several times before a vessel is in the desired shape. Although a long process, raising has the advantage of being able to produce shapes which are not necessarily round but can be used if need be to produce a complete sphere.

Whatever technique has been used, work-hardening will have been aggravated by damage.

Raising

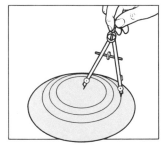

1 Draw circular pencil guidelines on the bowl.

2 Place smallest circle on the stake edge; hammer metal.

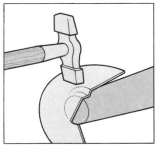

3 Turn work 12mm (½in.) for next blow. Repeat.

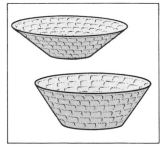

4 Raising should take place in gradual stages.

Removing the chains

The first job here is to remove the support chains. This must be carried out with care, so as not to distort the jump-ring. Take the ring in two sets of smooth-faced pliers and twist. The chains may now be slipped off the bowl, but first note each chain's position. The pans must now be thoroughly cleaned before any major amount of hammering takes place. Before doing so, however, you must consider whether the loss of any major area of patination will be detrimental to the value of the object. In the case of the scales the surface is clearly not good, having at some time been lacquered. This lacquer is now disintegrating.

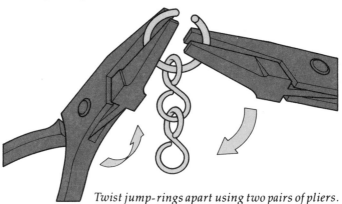

Twist jump-rings apart using two pairs of pliers.

Removing the lacquer

The lacquer can be removed by using a solvent and by working with a wet rag and pumice powder. The pumice powder should also remove most of the oxides on the inside surface of the pans. If it proves stubborn, immerse the pans in a pickling solution of approximately ten parts water to one part sulphuric acid. This should be in a broad-based 'Pyrex' type dish or plastic container preferably with a lid. It is also possible to purchase 'safety pickling' in small quantities for those people who do not wish to handle acids or have them around the house. As a useful addition copper and its alloys plus silver can be cleaned using citric acid (present in citrus fruits) and acetic acid (as found in vinegar) mixed with common salt. Dip a rag into the solution and swab the surface liberally. Whatever you choose make sure the object is well rinsed and dried before you move on to the next stage.

Removing dents

In order to remove the dent a bossing mallet is needed. This is a mallet that has an egg-shaped end. They can be made quite easily or a piece of wood with a domed end can be used. The work is held upon a sand bag or a wooden block which has a hollow gouged out of it. Sand bags are usually made of leather but an adequate substitute can be stitched up by using several layers of calico.

Hold the work firmly on a sand bag and mallet the bump from the inside. If the dent has a sharp crease line it will be necessary to push the surface out further than the original hollow. For sharp dents the use of a wooden punch is recommended.

The pan can now be reversed and placed over a mushroom-headed stake. Commercial stakes are very expensive but a good substitute can be made from a mild steel bar. For small work spare hammer heads held in the vice can be used to good effect. Stakes should be of slightly smaller radius than the work and need to have a very well finished surface. Any blemish on the stake will become impressed upon the inside of the pan. Using a cylindrical mallet, tap any raised bumps on the surface of the pan back flush with the surface. The pan should be manoeuvred over the stake head while the mallet is used to strike in the centre of the stake. The rim of the pan must also be tapped true. In order to keep forms regular it is important to tap gently around all the rim and not just where any imperfections are.

Removing dents

1 Push dents out from inside with the bossing mallet.

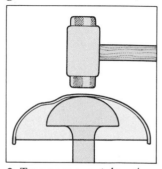

2 True up over a stake using a flat-faced mallet.

Planishing and polishing

We must now decide whether it is necessary to re-planish the form. Planishing has the effect of work hardening the form but it will also slightly flatten the form and thin the metal. If the metal has any severe creases, planishing will almost certainly be needed.

Planishing requires a special metal-faced hammer which is highly polished. As with the stakes, any imperfections in the hammer face will be imparted to the surface of the job. This is usually a finishing technique which will take out any ridges after hollowing or raising and will impart small regular facets to the surface.

Start by placing the dish on the stake and lightly tap the metal. Move the metal slightly until a clear ringing note is heard. This indicates that the metal is in contact with the stake below. Tap the metal, beginning at the centre, and

The repaired and polished scales are laid out ready for inspection and checking before final assembly.

Planishing

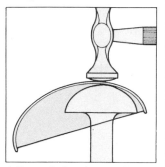

1 *Place bowl on the stake, and tap the metal.*

2 *Turn the bowl to produce concentric rings of facets.*

make sure the hammer strikes constantly over the same spot on the stake. Turn the metal to produce concentric rings of overlapping facets. Take care not to stretch the metal when getting near to the edge. The marks left after planishing should be slight and must not be confused with the coarse dimpling which appears on reproduction copperware at times. While planishing produces a finished surface, its primary use is to harden the piece and remove the raising or hollowing marks.

Providing the final planishing has been well done, the polishing of the surface should be relatively straight-forward if a polishing motor and mops are available. If not, the pieces may be polished by hand.

Initially any scratches that are not too deep can be removed by using a Water of Ayr (Scotch) stone (see p. 9). Make sure no flats or grooves are cut by the stone and always use plenty of water. There is also a range of jeweller's abrasive papers which are much less aggressive than emery cloth. The finest grades will produce a very satisfactory shine on their own. Final polishing can be done using jeweller's rouge and a chamois leather or soft cloth. Alternatively the rouge can be mixed into a paste and used with a stiff brush. For controlled polishing of individual areas either wrap a stick approximately $250 \times 25 \times 5$mm ($10 \times 1 \times \frac{1}{4}$in.) in felt or cloth and apply the rouge paste with this. Lastly clean off all traces of rouge with a degreasing agent. There are very good commercial metal polishes on the market which will easily maintain the shine you have achieved. If they are in use they will quickly obtain a pleasant patination

Hand polishing

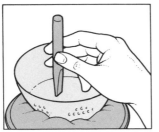

1 *Punch out sharp dents with a shaped wood tool.*

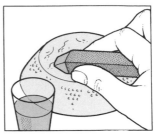

2 *Use Water of Ayr stone and water for scratches.*

3 *Brush in pumice and water, then rouge powder and oil.*

4 *Remove oxide with citric and acetic acids, and salt.*

but you may wish to lacquer the pieces at this stage. The lacquer will colour the metal and reduce its brilliance but against that you must balance the time you will spend cleaning.

Cleaning the chains

Now the bowls are finished, the chains can be removed from the balance arm and immersed in the cleaning solution. Leave them in the pickle for ten minutes, remove and wash thoroughly.

Take the chains and pin them to a piece of scrap wood. Now using a stiff brush, pumice and liquid soap, scrub them until all the oxides are removed, then rinse well. They should now have a fine matt finish. This will look quite acceptable but if you want a bright lustre, repeat the scrubbing using jeweller's rouge in paste form or a liquid polish. Chains can be buffed on a polishing motor but take care. They can be wrapped around a wooden strip and offered up to a small-diameter buffing wheel but make sure that there are no loose ends of chain or slack which can get snatched by the mop head of the machine.

The balance beam is iron and shows all the imperfections of being hand made. It has rust on it at present and this must be removed. On steel or iron items, brush off the surface dirt and try to assess the extent of cleaning required. Do not resort to the buffing wheel unless it is really necessary. There can be a real danger of destroying inscriptions, inlays, maker's marks, blueing, and so on.

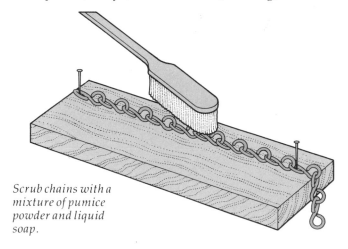

Scrub chains with a mixture of pumice powder and liquid soap.

Treatment for rust

There are basically two types of rust you will come across, the reddish brown, light surface coating and the more damaging hard blackish crust which pits the surface. Always use the minimum amount of abrasion required for its removal. A lot of effort can be saved by soaking the worst off by brushing in a paraffin (kerosene) and oil mix which is left to soak. There are a number of good rust removers on the market but make sure they are not used for too long otherwise they will etch the surface being cleaned. Always make sure you read the instructions as some rust removers have a surface priming action also. Abrasion of the rust can be carried out using steel wool, varying grades of emery cloth and, finally, polishing papers. If paraffin is used, make certain that any residue is rinsed away because otherwise it will cause rusting due to its high water content.

The finished surface can be lacquered or sealed using a wax polish. The balance can then be re-assembled, the bolt rings being held in two pliers once again and twisted closed. The balance is now ready for use or display.

Project 2: Policeman's lamp

The collecting of oil lamps has been popular for some time, although generally they have been of the domestic variety. With the recent upsurge of interest in all things from the past it has become increasingly difficult to find unspoilt examples of domestic lamps, consequently the collecting and display of commercial and industrial articles has become more widespread. This lamp is a good case in point. Articles connected with police work have always held a fascination for a small group of collectors. Now that interest has outgrown supply, all manner of police paraphernalia is being sought out. Things like helmets and painted truncheons have been fetching high prices on the open market.

The bull's-eye lamp

This lamp was made about 1880 in Bishopsgate, London, by Joyce and Son and still carries their brass plaque under the grime. It is a particularly nice piece as the bull's-eye lens is in very good condition and the body of the lamp itself is complete. Inside it carries the original burner which, strangely, has no adjustment. The wick would be set and lit at the beginning of the patrol and then left. The lamp gives off a very good beam of light. But there is a world of difference between sitting testing it in a darkened living room and trying to get it to penetrate the gloom of some fog-shrouded riverside alley. The lamp was designed to be either hand-held or clipped to the officer's belt. The bodywork is of tin plate and has been very well made as befits government-issue equipment. The component parts of the lamp have been cleverly designed so that the piece obtains maximum strength for little weight, the bull's-eye lens being set deep in the face of the lamp where it is well protected. The front casing is assembled around the lens and provides a watertight seal.

The general standard and skills exhibited in the original assembly of any piece should be carefully assessed. Any unusual technique should be noted and investigated further. The repair required for this lamp looks on first inspection to be a very simple affair. The chimney cap of the lamp is loose and must be removed, cleaned, trued up and resoldered.

Left: *The main components of the lamp were heavily contaminated with soot and grease. The lamp had to be taken apart* (top) *and all this removed before repairs could be carried out.*

Dismantling the lamp

Before attempting to unsolder the cap, it is necessary to take the central core of the lamp out of the casing. This is achieved by twisting the top cap and lifting. The burner, reflector and chimney sub-assembly should lift free of the main body. Now remove the burner casing and lift clear. If the burner is held by soot and grease, place it on top of a hot radiator until the grease becomes softened and allows the burner to be eased free. The top cap seems to be soldered and held on to the base cap with fixing tabs (see diagram below). By heating the area around the tabs the solder should melt and allow the tabs to be prised (pried) up with the end of an old screwdriver. Whilst doing this, support the lamp chimney assembly securely in a well-padded vice (vise). If, however, the sequence of making the lamp is considered, you will quickly realize that the tabs would have to be placed through fixing holes and soldered into position after the caps have been placed one on top of the other. This would mean that the tabs have also been fastened on the inside cap face (see diagram). On heating the areas concerned and removing the solder, the cap becomes looser but no nearer to coming free.

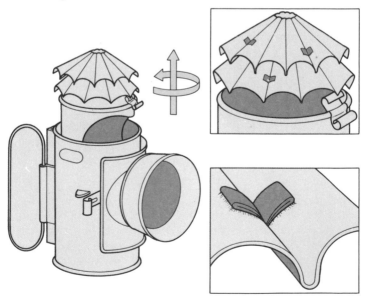

Remove the chimney from the body of the lamp by lifting and twisting (left). The top drawing shows the position of the soldered tabs, which are detailed above.

Releasing the cap

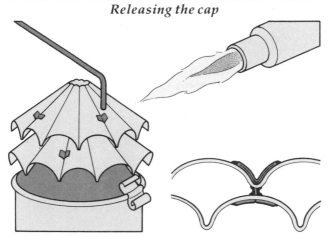

1 *Remove the tabs by heating the solder until liquid, and lifting with a metal pick.*

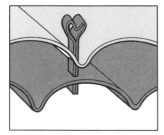

2 *Prise up the tab centre and unfold the ends.*

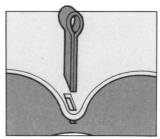

3 *The straightened tab can now be lifted free.*

On closer inspection of the fixing tabs they were found to be of a pattern which would prevent them releasing the cap even if they were very loose. The tabs were, in fact, hairpin-shaped. The two ends pushed through the caps and bent outwards. The loop left above the cap was pushed down in the middle, forming two narrow loops. These were then flattened and the cap sides and the whole assembly soldered.

Once this was discovered, it became comparatively easy to remove the cap. A large percentage of repairs call for the initial removal of component parts. It is essential that any would-be restorer tries to see the piece through the maker's eyes. Surprises will still occur, as in this case, but it should stop the frustration which may lead in turn to brute force and damage.

Cleaning

The lamp is now in several pieces: the top cap, the chimney, the burner and the lamp body. Most lamps of this age will be very dirty with dried oil and soot. Some will have heavy candle wax deposits inside.

If any solder repairs are required, the removal of grime is essential. First, place the parts of the lamp in a bowl of paraffin (kerosene). This part of the cleaning process can be speeded up by agitating the surface with a toothbrush or small brass scratch brush. When the majority of the dirt has been removed, transfer the parts to fresh paraffin to complete the process. Then wash the pieces in a solution of hot water, washing soda and a little washing-up (dishwashing) liquid. Take particular care in the cleaning of surface junctions; if necessary, clean these with a cotton-wool bud (Q-tip swab).

The article must now be thoroughly dried. The best way to do this is by placing all the pieces in a box of dry sawdust or alternatively, after initial drying, finish off with a hair drier.

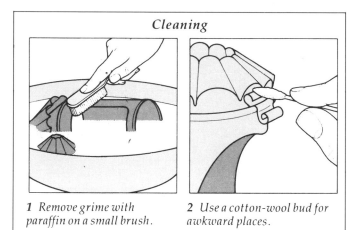

Cleaning

1 *Remove grime with paraffin on a small brush.*

2 *Use a cotton-wool bud for awkward places.*

Treating rust

If the surface is rusted, treat it with a proprietary rust remover. Choose one carefully – some commercial preparations leave a deposit which is a preparation for painting. Bear in mind, too, that any solutions left for long on the surface will etch the metal, possibly causing more damage than they cure. It might be possible to remove

Reaming with a wire drill

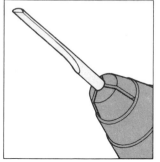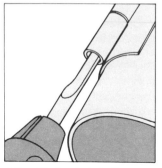

1 *File wire of a suitable gauge on one side.*

2 *Feed it slowly into the blocked hole of the handle.*

the rust with abrasives alone. Some rust lies just on the surface and can easily be removed with steel wool and a little oil. More stubborn areas can be tackled with emery cloth or a paste of emery grit, pumice and oil.

Apart from the cap there are two small problems to be taken care of. First, the handle on one side of the lamp is so stiff that any attempt to move it is liable to tear the metal. It can be eased, however, by treating the affected area with penetrating oil. Leave the oil to soak in for a time and then try to move the handle gently. It may require several applications to free the handle completely; you can then take the handle out and remove any remaining rust.

Rust inside the handle retaining hole can be cleaned off by reaming with a piece of wire held in a hand-drill. Take a piece of high tensile wire (piano wire) of a suitable gauge and file one side flat. Cut the end to approximately 30° and place in the drill. Turning the drill as you do so, push the wire into the hole. Move the drive gently forward, turning continuously. This will remove the rust without increasing the bore size.

The same solution is also used to eliminate the other problem, that of the hinge pin. This has become rusted and is causing the door of the lamp to sag badly. The pin must be removed, the hole reamed and the pin replaced with a larger wire. The replacement should ideally be of nickel, which is tough and will not rust.

Reaming using a wire drill is also very useful for hinges which do not have the knuckles exactly in line. The wire drill is used to ease the path of the hinge pin.

Trueing up the cap

With the piece now clean and rust-free, the cap can be secured in place. First, however, it must be trued up. This can be achieved by taking a wooden block into which you cut a tapered groove, which is then held in the vice. The cap is placed with one of its ribs in the groove. Now, using a narrow-faced mallet, tap the metal down into the groove; do this with all the ribs in a regular pattern. This way the cap will remain circular, whereas an attempt to re-form odd ribs is liable to distort the overall shape. The making of this cap is similar to a method of raising where the circumference of the disc is reduced by just this technique.

Making a new cap

If caps need to be replaced, they can be made by cutting a tin-plate disc and marking the number of ridges required. Form the first ridge with the mallet, then its opposite number, followed by the intervening ridges. The disc will rapidly start to make a cone-like shape. The ridges can then be trued up by using the grooved block and a shaped wooden punch.

The cap has been trued up using a shaped wooden block and wedge mallet.

Making a new cap

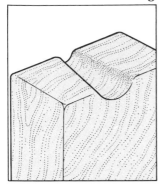
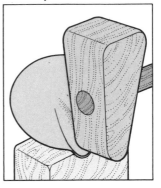

1 Use a wooden block with a tapered groove, and make the first flute with a narrow wedge-shaped mallet.

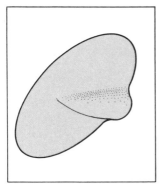
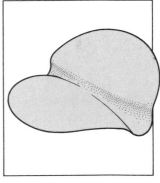

2 The flute runs from the rim to the disc centre.

3 Position the second flute opposite the first.

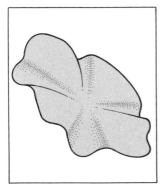
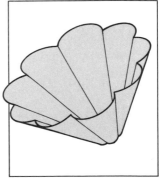

4 Place subsequent flutes opposite each other.

5 The disc gradually forms a cone.

Soft soldering

Once you are satisfied with the re-formed shape of the top cap, it can be resoldered on to the chimney. New retaining tabs can be cut from a sheet of tin plate. These tabs are approximately 25mm (1in.) long and 6mm (¼in.) wide. Bend them around a piece of dowel to form a hairpin shape. The cap is placed and the retaining tabs fed into position. Allow the tabs to pass through the lower cap by about 6mm (¼in.). Now use an orange stick to smear non-corrosive flux (see p. 9) on the tab ends.

Bend the tabs up underneath the lower cap, then gently heat the area and solder the tabs into position. The heating can be done with either a small flame from a gas torch or a soldering iron (the soldering iron is more convenient and safer). You can use either an electrically heated iron or one which requires preheating in a gas flame. If the latter type is used, heat until a green flame appears. This indicates that the iron is at the required temperature. The iron will need to be tinned before use. When it is hot enough, lightly file the surface to remove any oxides. Now dip the tip into the selected flux and touch on the solder. This should immediately coat the tip of the soldering iron. It is now tinned and ready to use.

Electric soldering irons are usually slimmer, as they have a constant heat source and do not rely on having a large hot mass to retain enough heat for the job. If a gas torch is used, it is important to ensure that over-heating does not take place. Most of these lamps were heavily tinned on the surfaces prior to painting and we do not want this to be damaged. If a flame is used, therefore, it must be kept constantly on the move. Once the tabs are soldered into position underneath, the top loop can be made in the upper tabs. The middle loop is pushed down with a screwdriver, forming two flatter loops. These are then burnished flat to the curved sides on the top of the cap. Paint these with flux and solder into position then remove any flux residue.

This particular lamp retains large patches of the original paint-work, but it also has areas of pitting by rust. You really need to decide for yourself whether to repaint and add a new surface to the objects in your care. There is no doubt, however, that an old damaged but original surface is valued higher in the auction rooms. Surfaces can nevertheless be treated with care and with consideration for patination and original colour without detracting too much from their value.

Fixing the cap

1 *Push new tab through the two caps.*

2 *Depress centre with a screwdriver.*

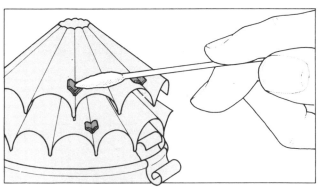

3 *Fold ends underneath the lower cap, and add flux to the top and bottom of tab.*

4 *Heat the area with a soldering iron.*

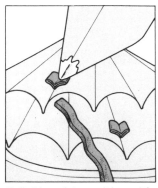

5 *Solder tab in position with soft solder.*

Project 3: Silver goblet

Although this item is relatively modern, the techniques and the problems involved in the repair of silver remain constant regardless of age.

The manufacture of silver has not changed for many hundreds of years. It is rather in the production and preparation of the raw materials that advances in technology have helped the silversmith. Silver has to be alloyed with other metal to make it hard enough for domestic use. By the end of the 12th century the alloying of silver and copper to the sterling standard was usual in the United Kingdom. Sterling silver is made up of 925 parts pure silver to 75 parts copper and this standard is still recognized in the UK.

The silver standard

Each item of silver made in the UK must conform to this standard and is tested at one of the assay offices situated around the country to make sure it does so. Scrapings are taken, analysed and if proved to be satisfactory, the piece is stamped (hallmarked). Each item must carry four stamps: a maker's mark, assay-office mark, standard mark and a year letter.

There is one other standard acceptable in the UK and that is of 958 parts per thousand silver and is known as Britannia silver. On no account should this be mixed with Britannia metal as this is pewter with a tin base. It was used briefly in the 17th century to discourage the melting down of the coinage. The standard was abolished in 1719 but can still be used for special items and receives a mark depicting the figure of Britannia.

In the United States hallmarking started in 1814 and manufacturers are personally responsible for marking their goods. The 'sterling' stamp is applied for 92.5% silver and 'coin' for 90%.

It is necessary to understand how important those marks are if you are undertaking any repair to silver. Any major replacement of material would invalidate the hallmark. In addition, there are many pieces which have a value out of all proportion to their material content because they bear the mark of a particular maker. It is vital that these marks are not erased or damaged.

If any repair is liable to interfere with the hallmark, obtain expert advice as to its effect.

Left: *The stem of this goblet was bent, and the base almost torn free – it had to be removed completely* (inset) *before repairs began.*

Repairs required

Repairs to this goblet, however, will not interfere in any way with the hallmarking. The goblet has suffered from bad packaging during moving house. The base has been squashed and has nearly been torn free of the item.

The piece is made in three parts, the bowl, the stem and the base. The bowl and the base have been made by spinning. This is a process in which a thin revolving disc of metal is shaped on a lathe over a metal or wooden former by using smooth, blunt tools. The disc which is in a softened state is held against the former and pressed down by the hand-held tool.

The stem has been cast and then soldered to both the bowl and the base. The joint at the point where the stem meets the base has been almost torn free, the base being distorted slightly in the process. The first job will be to separate the stem and base by heating with a gas torch until the solder flows, allowing the base to drop off.

The start of a delicate soldering job – the stem must be straightened and cleaned, then a new location found on the base.

Problems and precautions

Before the base and stem are separated, however, there are several things to be considered. We cannot be sure, if the item is heated overall, that the base will separate from the stem before the bowl does. This could be rather inconvenient as we will then have two repairs to do.

The goblet will have to be placed in the refractory material in such a manner as to deflect most of the heat from the bowl. On silver objects there is usually more than one grade of silver solder used. The temperatures at which these melt can vary between 630°C (1166°F) and 830°C (1526°F). Even at the lowest temperatures copper oxide or fire stain will form on the exposed surface. When heated, silver, which is an alloy, will separate into pure silver on the surface under which is a layer of oxide. On subsequent polishing the thin silver layer is stripped, leaving this much harder oxide which is seen as a grey shadow. A lot of time and effort is then needed to remove it, usually by pickling but if the layer is thick it has to be removed by abrasives.

In order to prevent the oxides forming it will be necessary to exclude the air from the surface. This can be done by painting the whole piece, inside and out, with a borax-based flux. There are also powders sold commercially which are specially designed to do this; one of these is Argo-Tec.

The Argo-Tec is mixed into a paste using methylated spirits (wood alcohol) and painted over the job. The spirits are ignited (make sure the surrounding area is safe) and driven off. This leaves a white powdery deposit which will melt and fuse on subsequent heating. If we were soldering, the job would be heated until the fusing of the Argo-Tec took place, then left to cool. Joint areas would be cleaned off, fluxed and then placed in their positions before the final soldering.

Using Argo-Tec

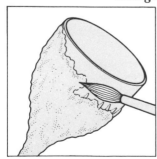

1 *Paint on Argo-Tec and methylated spirits.*

2 *Ignite meths and allow to burn out.*

Removing the base

For the present, heat the Argo-Tec until it fuses and then let it cool to a comfortable handling temperature. Make sure all the surface is covered, especially the edges, and place fire-bricks around the vessel to support it and to shield the cup. Heat up the base joint and gently pull off the base when the solder is flowing by using long tweezers or tongs. The surface coating is now removed by using boiling water.

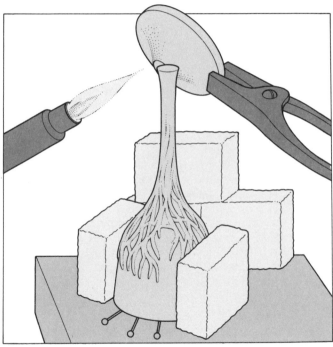

Heat solder joining stem and base, then remove base using tongs.

Repairing the base

Take the distorted base and, supporting it on the bench pin, file off any unwanted solder. The base at this stage can be carefully coaxed back into shape with a mallet and wooden former. Avoid using metal tools as they will stretch the material. Cut the profile of the base out of a piece of hard wood, and hold this in the vice (vise) as a support for the base plate while you are working on the base plate.

Repairing the base

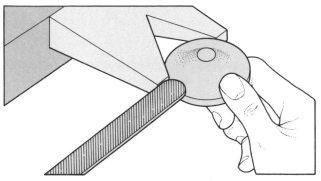

1 Support the base on the bench pin and file the stem contact area flat.

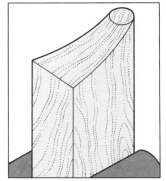

2 Wooden stakes support the base during remodelling.

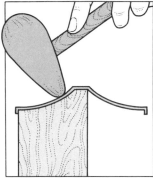

3 Correct distortions using a small mallet.

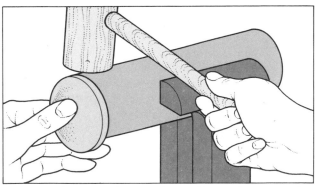

4 True up the rim on the edge of a metal bar with a boxwood mallet.

Repairing the stem

The base of the stem should also be filed free of old solder and generally cleaned up. Check that the stem is not bent. If any distortion has occurred, now is the time to correct it. To do so hold the vessel by the cup and support the stem on a wooden block. Turn the cup in your hand and watch the gap between the block and the stem. Any variation will indicate the amount of bend that must be removed. Continue revolving the cup and gently tap the stem onto the wood with a leather mallet. When you are satisfied that all is well, re-assembly can begin.

Mix fresh fire-stain inhibitor and liberally coat both parts of the vessel inside and out. Prepare as previously described and cool. Scrape the two areas where the base and stem touch and paint with flux. We will be using an extra easy grade of silver solder which melts at approximately 680-700°C (1256-1292°F). It is the lowest melting point of silver solder and should form a joint before the solder that is holding the cup to the stem melts.

Preparation for soldering

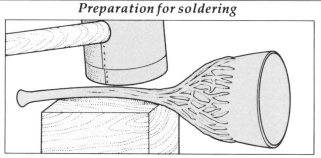

1 The stem can be straightened with a mallet. Rest stem on a wooden stake while you work.

2 Remove Argo-Tec from contact area with a file.

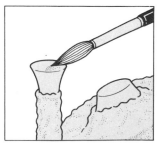

3 Apply fresh flux to both parts to be joined.

Wiring up

When soldering, one of the main problems is how to hold the components in place long enough to allow the joining to occur. For any pieces other than the smallest this can be achieved by using soft iron binding wire (the type used by florists in floral decoration). This wire is almost pure iron and is virtually free from spring. When heated it has a minimal amount of movement. As this is rather an awkward shape to wire, it is easier to use two small pieces of welding metal mesh. These will provide legs which the wire can be attached to and give the added benefit of raising the job clear of the fire-bricks. This will ensure that the base becomes heated quicker than if it was standing flat on the brick. The binding wires are cut to approximate length and a loop is made at each end. Next twist the wire about half-way along to provide a third loop. This will be used to tighten the wires. Now with the mesh in place top and bottom slip on the retaining wire, tightening each one using pliers. Make sure the goblet remains upright during this time and be careful not to over-tighten the wires. The object is to hold the goblet secure but if the wires are too tight, damage will occur as it tries to expand on heating. When you are satisfied that everything is as it should be, soldering can begin.

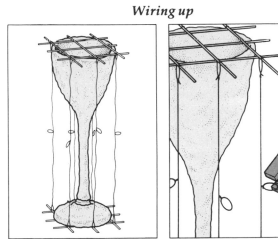

Wiring up

1 Use soft iron wire to hold pieces together.

2 Twist loops with pliers to tighten wires.

Soldering

Position the refractory material around the goblet so it will heat up quickly. The faster soldering occurs, once started, the better.

Begin with a soft, gentle flame, play it over the whole piece so that it warms up gradually and evenly. Any uneven heating may give rise to distortion. Watch the flux as it will bubble as the moisture is driven off, and make sure that this has not caused any movement in the joint. If everything is all right, increase the flame and bring the metal to the soldering temperature. Ideally

Soldering

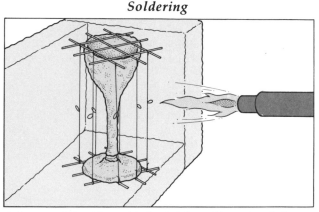

1 *Surround the object with fire-bricks and heat the whole piece with a gas torch.*

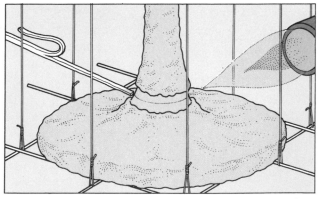

2 *When the flux flows as a liquid in the joint, introduce the solder.*

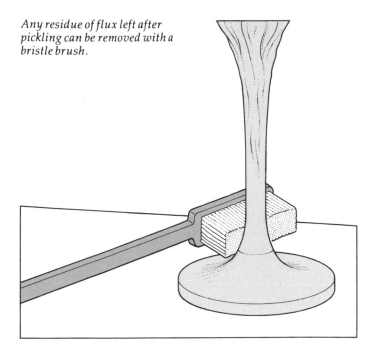

Any residue of flux left after pickling can be removed with a bristle brush.

soldering should be done in subdued light so that you will see the metal become a very dull, dark red. At this point the flux will flow as liquid – a very good indicator as to when to apply the solder. Feed the stick into the joint and it will 'flush' right around as a very bright silver streak. Reduce the flame immediately but do not withdraw it until you are satisfied that the solder has indeed flowed all the way round. Allow the vessel to cool and remove all the wires. Wash in very hot water, then place in the pickling solution to remove any remaining residues. Leave immersed for ten minutes then remove and wash thoroughly under running water. The areas of heavy texture can be brushed with a brass-bristled scratch brush. This should break up any particles of scale inhibitor remaining.

Polishing

It should now be possible to polish the piece using a commercial polishing cloth or liquid polish. Always try the polish on an easily cleaned area first and avoid polishes which leave a dark residue as this will be very awkward to remove from any crevices.

The repair is now complete.

Project 4: Copper kettle, trivet and burner

The base and burner

Copper's malleability – which makes it easy to form the material into almost any shape – and its ability to transfer heat, make it a natural choice for items like this kettle.

The piece we will be working on comes in four parts. First there is a trivet-like base, which supports the other components. This is made from perforated mild-steel plate held by wrought-iron 'c' scrolls. The overall condition of this piece is good to perfect and it should only require cleaning then refinishing as desired.

The lower part of the trivet holds the methylated-spirit (wood alcohol) burner. This is made from copper and brass and, apart from slight surface damage, is also in good condition. It has been machine-made but is of good quality because it has been made from a thick gauge of metal. Fortunately it has well-defined maker's registration marks and serial numbers in the base. These make it possible to date the piece very accurately.

The burner should require only slight cleaning in order to bring it back to its former condition.

The top plate

The third part of this set is placed on top of the trivet and acts as a hot plate for the kettle. When the burner is lit the heat would normally discolour a copper plate set over it but in this instance the top plate is double-skinned. As well as copper being decorative, it also holds heat very well – hence its use in soldering irons. Therefore this double skinning does not affect the heating of the kettle but prevents buckling of the top plate due to the direct application of heat. The top surface of the plate has been decorated by regular hammering which has caused concentric rings of dimples where the hammer has struck. This hammering, although primarily for decoration, also performs several other functions. The dimpling stiffens the sheet and helps prevent heat distortion. It also disguises surface damage due to the kettle sliding over it. Thirdly it helps prevent accidental movement of the kettle by creating a heavily textured surface. This top plate is very dirty underneath and will require a long soak to remove the oxides and grease, but there seems to be no major damage.

Left: *Although the kettle is covered in a light patination of small scratches and dents, there is one large dent* (top) *which must be removed before final cleaning and polishing.*

The kettle

The polished and restored kettle.

The last component of this set is the kettle which is made mainly from copper and copper alloys. All the parts have been machine-made apart from the spout of the kettle, which has been seamed and hand-formed.

The kettle handle is in direct contact with the body of the vessel and has no form of insulation. In use it must become extremely hot. On some sets of this kind, the kettle is held in a cradle which is tilted to pour water. This piece, however, has no such luxuries. The set is intended for use and not merely for decoration as the inside surfaces have been heavily tinned. The kettle is our major problem. Allowing for its age and the softness of the material the condition of the kettle appears to be good. There are, however, two large dents plus one small one on the same side. One of the dents is creased. The damage adds nothing to the general texture of the kettle and appears recent.

The handle has been riveted to the main body and is loose on one side. This will require tightening, or possibly replacing.

The base of the kettle is very lumpy and looks as

though it has been hit on many occasions, possibly in an attempt to remove swelling caused by the heat. With the type of burner used, good surface contact would be necessary to produce any significant rise in temperature in a reasonable time.

Removing the dents

With the dents positioned as they are, the main problem is one of access. While working on the scales (see p. 31), it was possible to hammer the dents out from the inside of the bowls. Here that is not the case.

There are two main ways of removing dents from an enclosed container: the level of difficulty determines the chosen method. How difficult it is will really be decided by the position of the dent in relation to the other features on the vessel. Any dent that is hard against the base or a rim wire will prove awkward, as will dents in spouts or at the junction of spout and body. All these positions prevent ease of access to the tools required. The shape of the dent is also of significance. If the impact has caused a sharp crease line to form, this always makes removal that much more difficult.

There is no sign of this unsightly dent on the restored item.

The snarling tool

If the dent cannot be reached with a hammer as in this case, it will be necessary to make up a snarling tool. This consists of a tapering bar which is cranked at both ends. One end is held tightly in the vice (vise) while the other end passes through the lid hole of the kettle and is cranked to an angle where it can touch the inside of the dent. The head of the tool should be polished and shaped to suit the dent being worked on. If the dent is sharp and narrow, file the head of the iron so it has a small area of impact. Keep the face of the snarling tool as smooth as possible because while it is pushing out the dent it will also be imparting a smooth internal finish. In the case of the kettle a well-finished surface will prevent damage to the tinned interior.

To use the iron hold it securely in the vice and rest the kettle on the other end. Move the kettle around until the head contacts the dent. Hold the kettle into your body with one hand and keep it as steady as possible. With your free hand holding an old hammer or piece of heavy wood strike the arm of the iron. This will cause the snarling iron to flex and rebound, striking the inside of the vessel. It is most important that it is held steady otherwise it will just get pushed out of the way and have little effect on the dent. Keep repeating this action until it has pushed the dent just past the surface, so there is a small lump rather than a dent. This lump is just the merest feature and should not be at all pronounced.

It is now possible to planish the area back level with the surrounding material. Sometimes the job can be rested on the head of the snarling iron when planishing. It must be remembered, however, that the iron has been especially made to vibrate so it is not very easy. If facilities for making a metal snarling tool are not available, it can be made from a piece of suitably springy timber. The same also applies to the stake on which to planish.

While working with a snarling tool as described above, it is possible to hold a piece of hard wood on the outside of the vessel against the point of contact inside. This stops the material getting pushed past the surface level and so minimizes planishing afterwards. Better results are achieved if, instead of wood, a piece of polished steel plate is used, because the quality of finish is improved.

If the area is to be planished, this must be carried out gently – the last thing that is wanted is excess stretching of the material and a consequent bulge.

Using the snarling iron

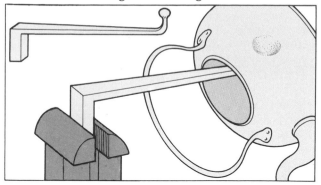

1 The snarling iron has a long springy neck. Set the kettle in position on the head of the tool.

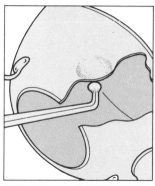

2 Position polished head of stake under dent.

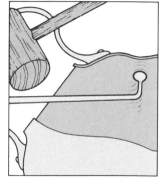

3 Strike iron so that vibrating head hits the dent.

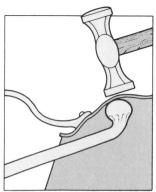

4 Material pushed clear can be planished later.

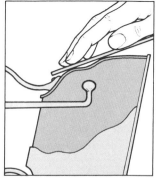

5 Polished steel behind the dent prevents excess raising.

Flattening the base

The base of the kettle should be tackled next. There are a large number of lumps and bumps to be removed and the main problem will be how to get to them. Initially they seem to be easy but the handle will get in the way of most of them. The excess material in the bulge must be eased away by stretching the surrounding metal. It is pointless just hammering the bulge as it will only appear somewhere else. However, we are not concerned with obtaining a dead flat surface on this particular object, merely to remove most of the unsightly bumps and obtain the maximum amount of contact with the hot plate that can reasonably be expected.

The method of doing this is to stretch the surrounding metal so that the bumps are pulled flat. Obviously there is a limit to how far the area around the bump can be stretched as it is defined by the 'dome' of the kettle.

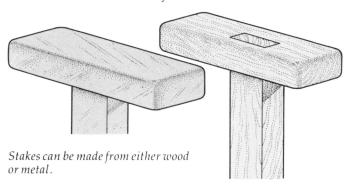

Stakes can be made from either wood or metal.

Using a stake

Before starting any hammer work several stakes will be required. These can be purchased but they are expensive if they will only be needed for the odd job; fortunately, suitable substitutes can be made. The stakes required need to be in the shape of a 'T' but with the top bar slightly off-set (see diagram above). It is usually possible to persuade the local garage or repair shop to weld a couple of pieces of scrap steel bar together for a small fee. Alternatively some home arc-welders will do the job very well.

A stake can, as a last resort, be made from a hardwood such as beech or maple. Care must be taken, however, in constructing the joint of the junction as it will have to take a lot of pressure.

Using a stake

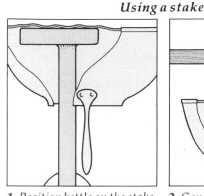

1 Position kettle on the stake with care.

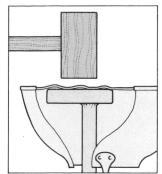

2 Gently ease bumps down with a mallet.

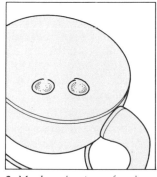

3 Mark perimeters of main areas of distortion.

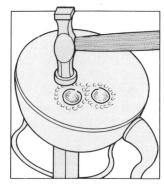

4 Gently pull bump edges flat with planishing hammer.

Set the stake in the vice and place the kettle on top. Now using a boxwood or rawhide mallet gently go over the base of the kettle to even out the bumps. It may be that this will be sufficient to give an acceptable surface. If not, the next stage is to identify the main areas of damage and mark these out with a felt pen. Anywhere on the surface within these boundaries is now left alone.

Using either a planishing hammer or one which you have shaped and polished, begin to tap gently around the perimeter of the areas marked. Ensure that the area being hammered is well supported on the stake or more harm than good will result. Continue with the planishing until the rim of the bumps becomes stretched and pulls the raised area flat. This can be encouraged with careful use of the mallet or wooden block.

Sharp indentations

In cases where there are small, sharp indentations it is often possible to push these out from the inside by using shaped wooden or metal punches and a hammer.

Place the object on which you are working on to a flat base which will give slightly under impact. This can be linoleum, a lead sheet or a wooden block with the end grain uppermost or anything that moves away from the point of contact. Support the punch just above the surface to be worked on, making sure that your hand is supported on the rim of the vessel or any convenient spot. Strike the punch with a light rhythmic action: the intention is to apply many rapid blows to the punch as you move it across the damaged surface. Watch the area of the job around where you are working and not the end of the punch which you are hitting. The force of the blow is controlled by keeping a close watch on the results of previous blows and appropriate force should be applied with action of the wrist rather than the arm. You will find just using the wrist less tiring and self-limiting in the amount of impact you can apply. You should be able to produce good results after a little practice.

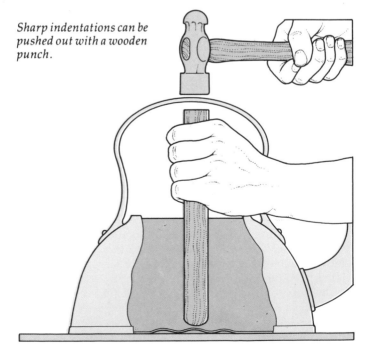

Sharp indentations can be pushed out with a wooden punch.

Using a wire

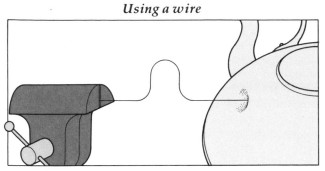

1 Solder the wire to the dent. Hold the other end of wire in the vice and pull sharply.

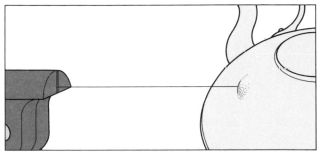

2 The wire will pull the dent proud of the surface. Several jerks may be needed.

Using a wire

There is an additional method of dent removal which can be used in extreme cases. It may not always be possible to get a snarling iron in contact with a dent because of its position. This is usually when it is hard up against a rim or protected by an overhanging feature. In this instance a wire is hard-soldered to the middle of the dent on the outside. The wire should be positioned so it is at right angles to the surface. The loose end of the wire is clamped firmly in a vice, while the object soldered to the other end is supported. Holding the vessel securely in both hands jerk the wire taut. This will pull the dent level. Several attempts might be needed before the level of the surrounding metal is reached. Ideally, the dent should be pulled just proud of the surface. The wire is then removed and the solder cleaned off with files. The area can be planished back to its original shape.

Air pressure used in planishing

It might not be possible to get a planishing stake behind the damaged area. In this case the vessel should be made airtight by putting on the lid and taping or closing the holes by using card and tape. The piece can now be held in the crook of your arm and *gently* planished using the air pressure inside for support.

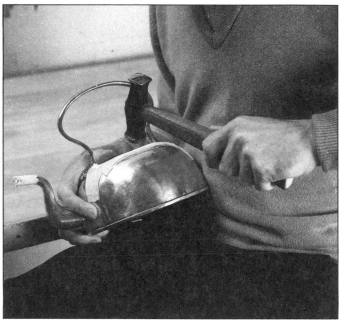

Tape the spout and lid, then hold the kettle firmly against your body to planish raised areas flush.

The handle

The handle of the kettle is rather loose on one side and ideally should be tightened. If it is not possible to make the rivet secure it will have to be drilled out and replaced. On this type of item that would be rather awkward as the rivet has been tinned on the inside of the kettle. This effectively has soldered it into place. Its removal would not be easy and of course the new rivet would need tinning if the kettle were to be used. As the inside head of the rivet is so secure there is a good chance of being able to expand the rivet on the outside so the handle is gripped.

Tightening the handle

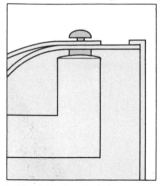

1 Position head of rivet firmly on the stake.

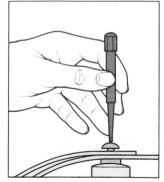

2 Using a centre punch strike rivet head.

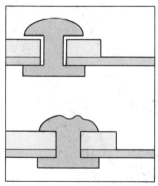

3 Expanding the rivet head takes up the slack.

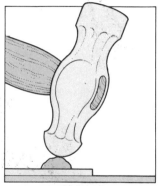

4 Use a ball peen hammer to re-form dome.

Place a cranked piece of metal that can support the kettle on the internal rivet head in the vice. Hold the kettle against your chest and place a tapering round-headed punch on the middle of the loose rivet. Make sure that your hand is well braced against the kettle. Strike the punch smartly so as to create a small dimple in the middle of the rivet. Hopefully, this will push the metal of the rivet outwards to grip the handle. Repeat as necessary. Now using a ball peen hammer gently re-form the domed rivet head, making sure that the kettle is still well supported and no damage is caused to the handle.

All that now remains is to clean and re-polish the surface of the kettle. Follow the advice given on pp. 33-5 for the best ways to do clean and polish copper.

Project 5: 19th-century percussion musket

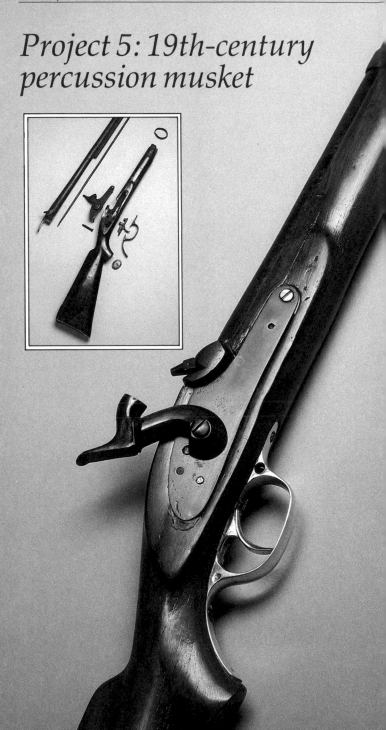

Restoring antique weapons

The repair and restoration of an antique weapon can cause more problems than most items. The main theme behind any work should be the restoration of the weapon to its original condition. It would not, however, be desirable to add or embellish any part or decoration which would not have initially been there.

Collectors, however, might not agree on the extent of repairs the amateur restorer should undertake. The individual skill of the restorer will of course play an important part in the decision as to what should or should not be tackled. With a muzzle-loading weapon such as the musket, a check should always be made to ensure it is unloaded. It is not at all unusual to find weapons with an undischarged load. When checking, the easiest method is to push a piece of dowel down the barrel and compare the length reached with the outside length of the barrel.

Should a charge be found, withdraw the load with great care and wash the barrel out with warm water. Dry it thoroughly. And *never* attempt to fire the musket, even when it is restored.

Stripping down

The first job will be to strip down the weapon. Remove the lock first, this can be achieved by undoing one or two screws which pass through the lock. If they are stubborn or rusty, placing a hot soldering iron on the end for a few minutes will usually loosen them.

The barrel can be removed by releasing the fixing screw in the tang and releasing any barrel bands. On some models retaining pins will also secure the barrel.

To remove the trigger assembly release retaining pins and extract any screws. As with many antique items much damage can be done by overcleaning. Armourer's marks and dates should be treated with great care.

By looking at things such as the retaining screws it is possible to judge whether the object has been repaired recently. Old screw threads are completely different from their modern counterparts, and usually far coarser in pitch. As the weapon is stripped down make notes so that you can re-assemble it in the correct order.

Left: *This musket is a major repair* (inset). *The lock is here returned to the partially restored stock.*

Dismantling the lock

Since the lock contains a number of springs, extra care is needed when taking it apart. Never release a screw until you are sure that the part it retains is relaxed. If you are at all unsure, place the lock in a plastic bag while releasing the screws. This will allow you to see what is happening but the bag will contain any sprung pieces which might fly off.

The particular musket pictured is in very bad condition. It will also be necessary to do a good deal of investigation about the piece as it appears to be made from several weapons which have been modified and married together. The lock is based upon that of a military musket but it has probably been manufactured in India. Possibly some pieces such as the lock plate are original and have had locally made parts added. The stock is based upon a European sporting arm and has had the military furniture such as the trigger guard modified to suit. To restore this piece completely would involve a lot of repair to the wooden stock.

As this will be such a major repair, the following section will deal with specific jobs which would be likely to be found when dealing with antique weapons.

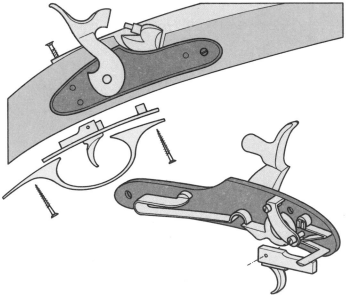

To remove the trigger assembly, first remove the guard retaining screws. Take care when releasing the springs of the lock.

Trigger guard

The end of the guard has snapped off. This can be repaired by hard soldering on a new piece. The main problem will be trying to hold the guard steady during soldering. For this reason it will be easier if the piece to be added is left over-size to allow for any movement.

Clean up both ends to be soldered and make sure they are filed square and well fitting. Flux each end and support both pieces on fire-bricks. Keep them clear of the surface by laying them on split pins which have been opened out to a V-shape. Heat both parts, but remember to build up the heat in the large piece first. When the flux fuses apply the solder. It will be better to use a stick rather than a small chip. Sometimes if heating and subsequent soldering is not immediate, the solder can be bled from the chip leaving the skin or husk remaining. Any attempt to continue heating in the hope of melting the husk and getting it to flow usually ends up with over-heating the job and a depression occurs on the surface where the chip was.

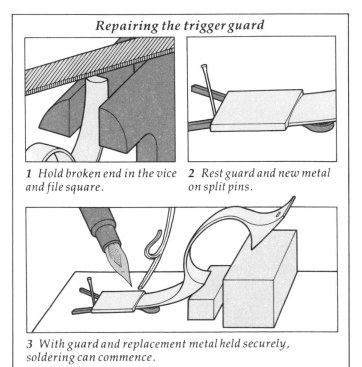

Repairing the trigger guard

1 Hold broken end in the vice and file square.

2 Rest guard and new metal on split pins.

3 With guard and replacement metal held securely, soldering can commence.

Finishing the trigger guard

After cleaning off the flux residue and oxide, the trigger guard can be filed to shape. Any adjustment to the curved form can be done at this stage by tapping the guard with a mallet while supporting it over a steel bar. Final finishing and polishing can be achieved with abrasive papers and rouge.

Forming the guard

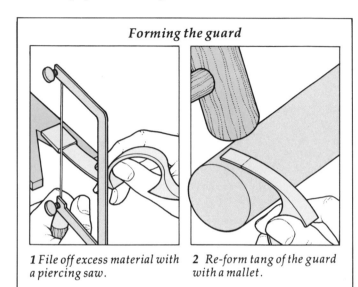

1 File off excess material with a piercing saw.

2 Re-form tang of the guard with a mallet.

Barrel ramrod pipe

The barrel of this weapon is very heavily pitted especially around the area of the nipple. This is only to be expected due to the corrosive qualities of the firing charge. The nipple needs to be removed but this should only be attempted following prolonged soaking in a penetrating oil. Even then it should be undertaken with care to prevent the screw shearing off in the breedle plug.

Before any further action is taken the heavy rust on the ramrod pipe and barrel needs to be killed. This can be done using a proprietary rust remover. Make sure that the muzzle is blocked and also the nipple hole is plugged before applying the anti-rust agent.

Further cleaning of the barrel can be undertaken with steel wool and oil or, if necessary, by using emery cloth. Because the barrel is so badly corroded, it would be unwise to try and reach a finish which would enable reblueing to take place as this will require a lot of material removal to obtain the high polish required. Providing the rust has stopped and any loose material has been removed, the surface will look satisfactory if it is finished with fine steel wool and beeswax. This can be burnished to a high lustre.

Before that takes place, however, there are several large dents in the ramrod pipe which must be removed.

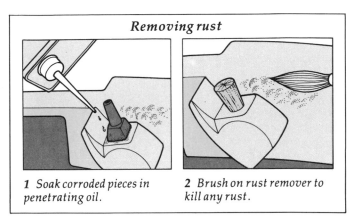

Removing rust

1 Soak corroded pieces in penetrating oil.

2 Brush on rust remover to kill any rust.

Take a steel rod at least three quarters the pipe length and fractionally less than the internal diameter. This should have one rounded end. Take the rod and gently tap it into the pipe in order to push out the dent. Do not hammer too hard as there is, of course, a danger of getting it jammed. Tap the rod in a little at a time and then withdraw. Grease pushed into the bore will assist its withdrawal. Keep moving forward each time you enter and gradually ease the depression out. If the walls of the pipe are too resistant it might be necessary to heat them.

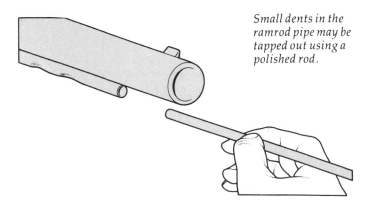

Small dents in the ramrod pipe may be tapped out using a polished rod.

Heating the barrel

Place the barrel in a vice (vise) and, using a gas torch, heat up the ramrod pipe at the dent. As the pipe has probably been soldered to the barrel you will have to prevent the rest of the pipe from getting hot. This can be done by wrapping the remaining area in wet bandages. When the dent area is hot, try tapping the rod into the pipe. This will need to be undertaken quite quickly as the rod can become trapped by the cooling metal. When all dents have been satisfactorily removed complete the final finishing of the barrel.

If using heat to remove dents, keep heat localized.

Hammer

The hammer on this lock is also very badly corroded. The striking face of the hammer has been virtually eaten away over the years. The hammer tail has been worn smooth and shows hardly any chequering marks.

The body of the hammer has been crudely modified to suit the weapon. It has been filed very thin and exhibits nothing of the full, well-rounded form of its contemporaries. This has been done so that the hammer could be riveted in place rather than held by a screw. It will be necessary to cut the rivet in order to free the hammer. Once this has been done any surface corrosion must be removed either chemically or mechanically.

Use a home arc-welder to deposit fresh material around the nose and flank of the hammer. (Alternatively, ask a local workshop to do this for you.) Then, file back the excess to form the desired profile. It may be that this has to occur several times before a satisfactory surface is

Repairing the hammer

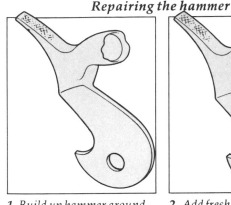

1 Build up hammer around the head and body.

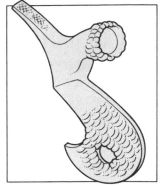

2 Add fresh material to the hammer by welding.

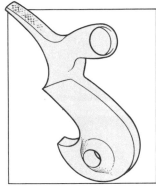

3 File new material to the correct profile.

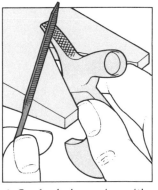

4 Cut fresh chequering with a knife edge needle file.

achieved. The recess for the nipple can be drilled out by using a very shallow ground drill bit, the base of the hole being almost flat. The chequering on the tail can be cut using a three square or knife edge needle file. There are special files for chequering but they do tend to be expensive. As the hammer is in such a bad state it is important that research into hammer profiles is carried out before final assembly of the lock.

The lock itself must be cleaned to remove any corrosion and the parts which interact polished to a high finish.

Remember that it is far better to have an item which, when clean, still exhibits elements of its age and history rather than being over-cleaned and lacking any period feel or crispness.

Acknowledgements

Swallow Books gratefully acknowledge the assistance given to them in the production of *Care and Repair of Antique Metalware* by the following people and organizations. We apologize to anyone we may have omitted to mention.

Photographs: Jon Bouchier 4, 6, 8, 9, 10, 11, 13, 26, 32, 36, 46, 48, 56, 58, 59, 66, 68; The Bridgeman Art Library 7.

Illustrations: Graham Bingham 15, 19, 20, 21; John Woodcock 28, 29, 30, 31, 33, 34, 35, 38, 39, 40, 41, 42, 43, 45, 49, 50, 51, 52, 53, 54, 55, 61, 62, 63, 64, 65, 67, 70, 71, 72, 73, 74, 75.

Tools and materials on pages 9-13 supplied by H. S. Walsh & Sons Ltd (incorporating Charles Cooper, Hatton Garden).